F-35 Joint Strike Fighter (JSF)

Executive Summary

- Flight test teams operating the 18 test aircraft assigned to the developmental flight test centers nearly matched or exceeded flight test sortie goals through October 2013. This occurred despite loss of several government employee work days due to furloughs and sequestration, and two fleet-wide grounding instances. Flight sciences testing made the planned progress in envelope expansion and handling qualities for the year; however, mission systems and weapons integration testing made little progress and continued to fall behind test point execution goals driven by upcoming fleet release and Services' Initial Operational Capability plans.
- Mission systems development and test teams focused on getting Block 2B capability into flight test, which began several months later than planned in the integrated master schedule. Block 2B capability is the next major increment planned to be released to the fleet of production aircraft, and the first planned to have combat capability. A considerable amount of testing was necessarily devoted to completing development of prior-block capabilities, attempting to complete fixes to known problems, and regression testing of new versions of software. As a result, through October 2013, little progress was made in completing flight testing required by the baseline Block 2B joint test plan. This creates significant pressure on development and flight test of the remaining increments of Block 2B, with approximately 12 months remaining on the program timeline before final preparations are planned to begin for an operational utility evaluation of the combat effectiveness and suitability of Block 2B.
- Weapons integration, which includes both flight sciences and mission systems test events, did not make the planned progress in CY13. Weapons integration is recognized by the program as a critical path to both Block 2B completion and the end of Block 3F development.
- Flight operations of production aircraft and upcoming operational testing of Block 2B capability depend on the functionality of the Autonomic Logistics Information System (ALIS), which has been fielded with significant deficiencies. The current ALIS capability forces maintenance operations into numerous workarounds and causes delays in determining aircraft status and conducting maintenance. The program expects improvements in the next ALIS version, scheduled in time for the release of Block 2B capability to the fleet, but there is no margin in the development and test schedule.
- F-35B flight test aircraft completed 10 days of testing aboard USS *Wasp* as planned in August 2013. Testing included evaluating changes to control laws, expanding the operational flight envelope, and flight operations at night.
- Overall suitability performance continues to be immature, and relies heavily on contractor support and workarounds unacceptable for combat operations. Aircraft availability

and measures of reliability and maintainability are all below program target values for the current stage of development.
- The program is now at significant risk of failing to mature the Verification Simulation (VSim) and failing to adequately verify and validate that it will faithfully represent the performance of the F-35 in the mission scenarios for which the simulation is to be used in operational testing.
- The program completed F135 engine vulnerability test series that demonstrated:
  - The engine can tolerate a range of fuel leak rates ingested through the inlet to simulate and assess ballistically induced fuel tank damage effects. System-level live fire tests using a structural F-35C test article with an operating engine will determine the engine tolerance to the fuel quantity ingested as a result of actual ballistic damage.
  - The engine is tolerant of mechanical component damage from single-missile fragments, while fluid-filled engine components are vulnerable to fire. Results from two tests demonstrated engine vulnerabilities against more severe threats and were consistent with results from prior legacy engine tests.
- The program examined the F-35 vulnerability to ballistically induced damage to the F-35 gun ammunition. Missile fragment ballistic testing on single PGU-32 rounds demonstrated that a propellant explosive reaction and sympathetic reaction of adjacent rounds in multiple round tests were unlikely. The F-35 is, however, vulnerable to ballistically-induced propellant fire from all combat threats.
- The vulnerability of the F-35 to electrical system ballistic damage remains an open question. Based on the F-35A aircraft (AA:0001) in-flight incident in 2007, electrical arcing

tests in 2009, and the flight-critical system-level test events in 2012, DOT&E recommended that the program conduct additional analyses to address the likelihood and consequence of arcing from the 270-volt to 28-volt system. The Lockheed Martin electrical power system team is currently working on a response to these concerns.
- The program provided no update on the decision to reinstate the Polyalphaolefin (PAO) shut-off valve, a 2-pound vulnerability reduction system that could reduce crew casualties and the overall F-35 vulnerability by approximately 12 percent, averaged across all threats and F-35 variants.
- The program redesigned the On-Board Inert Gas Generation System (OBIGGS) to meet vulnerability reduction and lightning requirements. The program is currently planning the tests for FY14 to ensure that the system is able to maintain fuel tank inerting throughout all mission profiles. The system should protect the F-35 from threat-induced or lightning-induced fuel tank explosions.

Actual versus Planned Test Metrics through October 2013

TEST FLIGHTS					
	All Testing	Flight Sciences			Mission Systems
	All Variants	F-35B Only	F-35A Only	F-35C Only	
2013 Actual	993	284	226	181	302
2013 Planned	985	287	241	171	286
Difference from Planned	+0.8%	-1.0%	-6.2%	+5.8%	+5.6%
Cumulative Actual	3,601	1,269	963	612	757
Cumulative Planned	3,284	1,127	910	584	663
Difference from Planned	+9.7%	+12.6%	+5.8%	+4.8%	+14.2%

TEST POINTS									
	All Testing	Flight Sciences			Mission Systems				
	All Variants	F-35B Only	F-35A Only	F-35C Only	Block 1*	Block 2A	Block 2B	Block 3	Other
2013 Baseline Accomplished	5,464	1,418	1,713	1,032	326	168	461	0	346
2013 Baseline Planned	7,180	1,701	1,836	1,165	1,755			0	723
Difference from Planned	-23.9%	-16.6%	-6.7%	-11.4%	-45.6%				-52.1%
Added Points	1,776	178	193	211	1,194			0	0
Points from Future Year Plans	720	320	0	400	0			0	0
Total Points Accomplished**	7,960	1,916	1,906	1,643	2,149			0	346
Cumulative SDD Baseline Actual	26,689	9,356	7,636	5,859	1,166	614	860	0	1,198
Cumulative SDD Baseline Planned	27,075	9,256	7,735	5,564	2,663			0	1,857
Difference from Planned	-1.4%	+1.1%	-1.3%	+5.3%	-0.9%			0.0%	-35.5%
Program Office Estimated Test Points Remaining	31,218	9,726	6,057	7,493	350	606	3,226	1,739	2,021

* Includes Block 0.5 and Block 1 quantities
** Total Points Accomplished = 2013 Baseline Accomplished + Added Points
SDD = System Development and Demonstration

System
- The F-35 Joint Strike Fighter (JSF) program is a tri-Service, multi-national, single seat, single-engine family of strike aircraft consisting of three variants:
  - F-35A Conventional Take-Off and Landing (CTOL)
  - F-35B Short Take-Off/Vertical-Landing (STOVL)
  - F-35C Aircraft Carrier Variant (CV)
- It is designed to survive in an advanced threat (year 2012 and beyond) environment using numerous advanced capabilities. It is also designed to have improved lethality in this environment compared to legacy multi-role aircraft.
- Using an Active Electronically Scanned Array radar and other sensors, the F-35 is intended to employ precision-guided bombs such as the Joint Direct Attack Munition (JDAM) and Joint Standoff Weapon, AIM-120C radar-guided Advanced Medium-Range Air-to-Air Missile, and AIM-9 infrared-guided short-range air-to-air missile.
- The program provides mission capability in three increments:
  - Block 1 (initial training)
  - Block 2 (advanced training and initial combat)
  - Block 3 (full combat)
- The F-35 is under development by a partnership of countries: the United States, Great Britain, Italy, the Netherlands, Turkey, Canada, Australia, Denmark, and Norway.

F-35 JSF

Mission

- A force equipped with F-35 units should permit the Combatant Commander to attack targets day or night, in all weather, and in highly-defended areas of joint operations.
- F-35 will be used to attack fixed and mobile land targets, enemy surface units at-sea, and air threats, including advanced cruise missiles.

Major Contractor

Lockheed Martin, Aeronautics Division – Fort Worth, Texas

Test Strategy, Planning, and Resourcing

- The JSF Program Office, in coordination with the Services and the operational test agencies, submitted Revision 4 of the Test and Evaluation Master Plan (TEMP) for approval in late CY12.
    - DOT&E approved the TEMP in March 2013, under the condition that the schedule in the TEMP be revised such that no overlap exists between the final preparation period for IOT&E and the certification period required for the Services' airworthiness authorities to issue flight clearances.
    - DOT&E required that the final preparation for the IOT&E could not begin any earlier than the Operational Test Readiness Review, a point in time when the JSF Program Executive Officer certifies the system ready for IOT&E.
- This report reviews the program by analyzing the progress of testing and the capability delivered as a function of test results. The program plans a specific set of test points (discrete measurements of performance under specific test conditions) for accomplishment in a given calendar year. In this report, test points planned for a given calendar year are referred to as baseline test points. In addition to baseline test points, the program accomplishes test points added for discovery, regression of new software, and regression of fixes to deficiencies identified in flight test. Cumulative System Development and Demonstration (SDD) test point data refer to the total progress towards completing development at the end of SDD.

F-35A Flight Sciences

Flight Test Activity with AF-1, AF-2, and AF-4 Test Aircraft

- F-35A flight sciences testing focused on:
    - Accomplishing clean-wing (no external stores or weapons) flutter testing of the full Block 2B flight envelope with weapons bay doors closed and open
    - Evaluating flying qualities with internal stores (GBU-31 JDAM, GBU-12 laser-guided Bomb, and AIM-120 Advanced Medium-Range Air-to-Air Missile) and external stores (AIM-9X short-range missile)
    - Characterizing the subsonic and supersonic weapons bay door and environment
    - High angle-of-attack (above 20 degrees) testing in clean configuration and in landing configuration
- F-35A flight testing was affected by two directives to halt testing in early CY13.

- The entire F-35 fleet was grounded on February 21, 2013, after a crack was discovered on February 19, 2013, in one of the third-stage, low-pressure turbine blades in the engine of AF-2, a flight sciences test aircraft at Edwards AFB, California. The cause of the crack was determined to be a rupture due to thermal creep, a condition in which deformation of material forms from the accumulated exposure to elevated temperatures at high-stress conditions. The stop order was lifted one week later, on February 28, 2013, with the requirement for additional inspections of the engines to ensure the effects of creep, if they occur, are within tolerances.
    - Discovery of excessive wear on the rudder hinge attachments on AF-2 in early March 2013 also affected availability of test aircraft. As a result, the test fleet was grounded for inspections and maintenance actions, including replacing part of the hinge on AF-2 and adding wear-preventing washers to the hinges of the rest of the test fleet.
    - In total, AF-2 was down for six weeks for replacement of the engine and rudder hinge repair.
- The test team completed supersonic clean wing flutter testing with the weapons bay doors open and closed, clearing the F-35A Block 2B envelope to 1.6 Mach/700 knots calibrated airspeed.
- The team began testing F-35A controllability at high angles of attack and high yaw rates, including the first intentional departures from controlled flight with external stores.
- The test team completed all weapons safe-separation events of GBU-31, JDAM, and AIM-120 weapons for the Block 2B envelope by the end of August. These tests precede end-to-end weapons delivery accuracy test events performed with mission systems test aircraft.
- The program tested two aircraft modified with new horizontal tail surface coatings and instrumented with temperature sensors to monitor heating from conditions of extended afterburner use. Damage to horizontal tail coatings was previously discovered during flight tests on all three variants involving extended use of the afterburner not expected to be representative of operational use, but which was necessary to achieve certain test points. Non-instrumented test aircraft continue to operate with restrictions to the flight envelope and use of the afterburner.

Flight Sciences Assessment

- Through the end of October, the F-35A flight sciences test team lagged in completing the planned flights for the year, having accomplished 226 sorties against the plan of 241. Productivity in baseline test points also lagged by 6.7 percent, as the team accomplished 1,713 baseline points against a plan of 1,836.
- The amount of added work from new discoveries or from regression of new versions of air vehicle software (i.e., control laws governing performance and handling qualities) has been less than expected through the end of October. The team allocated 311 points for growth, but accumulated only 193 growth test points by the end of October.
- The test team accomplished test points for clearing the flight envelopes for Blocks 2B and 3F.
  - Progress through the Block 2B test points was accomplished according to the plan, with 1,089 Block 2B points accomplished compared to 1,083 planned.
  - The team also accomplished test points needed to clear the Block 3F flight envelope, but did so at a rate behind the plan. Through the end of October, the team accomplished 624 Block 3F envelope test points against the plan of 753 points, or 83 percent of the plan. The work accomplished for the Block 3F envelope included points with weapons bay doors open and with external air-to-air weapon load-outs.
- Weight management of the F-35A variant is important for meeting air vehicle performance requirements. Monthly aircraft weight status reports produced by the program compute a sum of measured weights of components or subassemblies, calculated weights from approved design drawings released for build, and engineering weight estimates of remaining components.
  - According to these reports, the weight estimates for the F-35A decreased by 72 pounds from January to October 2013. The latest October 2013 F-35A weight status report showed the estimated weight of 29,030 pounds to be within 341 pounds of the projected maximum weight needed to meet the technical performance required per contract specifications in January 2015.
  - Although the weight management of the F-35A has demonstrated a positive trend over the past year, this small margin allows for only 1.16 percent weight growth over the next year to meet contract specification requirements in January 2015. The program will need to continue rigorous weight management beyond the contract specification timeline endpoint in January 2015 and through the end of SDD to avoid performance degradation and operational impacts.
- F-35A discoveries included:
  - During early high angle-of-attack testing, problems with the air data computer algorithms were discovered, requiring an adjustment to the control laws in the air vehicle software and delaying a portion of the testing until the updated software was delivered to flight test in September. High angle-of-attack testing resumed, and is required to support the full flight envelope and weapons employment capabilities planned for Block 2B.
  - Buffet and transonic roll-off (TRO) continue to be a concern to achieving operational capability for all variants. The program changed the flight control laws to reduce buffet and TRO in the F-35A. No further changes to the control laws are being considered, as further changes will potentially adversely affect combat maneuverability or unacceptably increase accelerative loading on the aircraft's structure. The program plans to assess the operational effect of the remaining TRO and the effect of buffet on helmet-mounted display utility by conducting test missions with operational scenarios in late CY13 and early CY14.

F-35B Flight Sciences

Flight Test Activity with BF-1, BF-2, BF-3, BF-4, and BF-5 Test Aircraft

- F-35B flight sciences focused on:
  - Continued expansion of the Block 2B flight envelope
  - Expansion of the envelope for vertical-lift and short take-off operations, including operations with external stores and the gun pod (mounted on the centerline station)
  - Flight clearance requirements for the second set of ship trials on the USS Wasp
  - Block 2B weapons separation testing (for GBU-12, GBU-32, and the AIM-120 missile)
  - Fuel dump operations with a redesigned dump valve and flap seals
  - Initiating high angle-of-attack testing
  - Completing tanker air refueling with strategic tankers, i.e., KC-135 and KC-10 aircraft
  - Regression testing of new vehicle systems software
- The F-35B fleet was grounded after the first British production aircraft, BK-1, experienced a fueldraulic line failure in the STOVL-unique swivel nozzle at Eglin AFB, Florida, on January 16, 2013. The cause was determined to be a poor manufacturing process used for the hoses, leading to crimping dimensions being out of specification; the stop order was lifted nearly four weeks later on February 11, 2013, allowing all F-35B flights to resume.
- The program modified one F-35B test aircraft with new coatings on the horizontal tail to address deficiencies seen in bonding of the skin under high-temperature and high-airspeed conditions. These conditions involve extended use of the afterburner not expected to be representative of operational use but which was necessary to achieve certain test points. The new bonded coating failed during flight test and experienced dis-bonding and peeling. The program continues to investigate the effects of afterburner use on the horizontal tails and plans to modify two F-35B test aircraft with new coatings and temperature sensing instrumentation to collect more data. Non-instrumented test aircraft continue to operate with restrictions to the flight envelope and use of the afterburner.

Flight Sciences Assessment

- Through the end of October, the F-35B flight sciences test team accomplished 284 of 287 planned flights, a shortfall of 1 percent. Completion of baseline test points was short by nearly 17 percent, as the team accomplished 1,418 of 1,701 planned baseline points. Similar to the F-35A flight science testing, the amount of added points due to growth was lower than expected, as the team flew only 178 growth points through the end of October, below the 287 points planned.
- Completed workup and second set of ship trials (referred to as DT-2) on time. The primary objective of the test period was to collect data for providing a ship-based flight envelope for vertical landings and short take-offs to support Block 2B fleet release and Marine Corps Initial Operational Capability. Flight activity included night operations and inert internal weapons stores.

- Progress through weapons safe-separation testing was behind the planned schedule, as only 12 of the planned 22 separations had been accomplished.
- Progress through the work needed to release the Block 2B flight envelope also lagged the plan, with completion of 1,247 of the 1,530 baseline points. Some weapons-related points were blocked earlier in the year when a problem with the GBU-12 lanyard was discovered, requiring a new lanyard and procedures to be developed. The test team was able to accomplish additional points in the Block 3F envelope – similar to the work being done in the F-35A flight sciences – completing 491 points against the plan of 171, pulling forward 320 points from future Block 3F test plans.
- The following table, first displayed in the FY11 Annual Report, describes the observed door and propulsion problems by component and identifies the production cut-in, if known.

F-35B DOOR AND PROPULSION PROBLEMS				
Category	Component	Problem	Design Fix and Test Status	Production Cut-In
Structure	Auxiliary Air Inlet Door (AAID)	Inadequate life on door locks, excessive wear and fatigue due to the buffet environment, inadequate seal design.	New designed doors are being installed on low-rate initial production (LRIP) aircraft as part of the ongoing modification plan; five completed through the end of September. Fatigue testing started in November 2012 and has completed just over 6 percent of the planned two lifetimes of testing as of end of September.	BF-38 LRIP 6 2014
Propulsion	Drive Shaft	Lift fan drive shaft undergoing a second redesign. Original design was inadequate due to shaft stretch requirements to accommodate thermal growth, tolerances, and maneuver deflections. First redesign failed qualification testing.	New design of the drive shaft will begin qualification testing in December. Full envelope requirements are currently being met on production aircraft with an interim design solution using spacers to lengthen the early production drive shaft.	BF-50 LRIP 8 2016
Propulsion	Clutch	Lift fan clutch has experienced higher than expected drag heating during conventional (up and away) flight during early testing.	New clutch plate design, with more heat-tolerant material, is complete. Clutch plates are being thinned on LRIP 5 and 6 aircraft, at the expense of reduced life (engagements) to the clutch, to prevent drag heating.	BF-44 LRIP 7 2015
Propulsion	Roll Post Nozzle Actuator	Roll post nozzle bay temperatures exceed current actuator capability; insulation is needed to prevent possible actuator failure during vertical lift operations.	Insulation between the roll post nozzle bay and the actuators is being installed in pre-LRIP 7 aircraft to allow unrestricted operations; however, the actuators must be replaced at 1,000-hour intervals. New actuators will be installed in LRIP 7 aircraft and beyond, removing the requirements for the insulation and extending the service life to 4,000 hours.	BF-44 LRIP 7 2015
Propulsion	Bleed Air Leak Detectors	Nuisance overheat warnings to the pilot are generated because of poor temperature sensitivity of the sensors; overheats are designed to be triggered at 460 degrees F, but have been annunciated as low as 340 degrees F.	More stringent acceptance test procedures are in place, requiring the sensors to be more accurate. Maintenance personnel are checking the detectors on pre-LRIP 5 aircraft, and replacing them in accordance with directives, if necessary.	BF-35 LRIP 5 2014
Propulsion	Aux Air Inlet Door Aft down-lock seal doors (aka "saloon doors")	Doors are spring-loaded to the closed position and designed as overlapping doors with a 0.5-inch gap. The gap induces air flow disturbance and make the doors prone to damage and out-of-sequence closing. Damage observed on flight test aircraft.	Springs are being limited to 4,000 hours or half the planned lifetime. Program continues to investigate whether a new design to the doors is required.	TBD

- Weight management of the F-35B aircraft is critical to meeting the Key Performance Parameters (KPPs) in the Operational Requirements Document (ORD), including the vertical lift bring-back requirement. This KPP requires the F-35B to be able to fly an operationally representative profile

and recover to the ship with the necessary fuel and balance of unexpended weapons (two 1,000-pound bombs and two AIM-120 missiles) to safely conduct a vertical landing.
  - Weight reports for the F-35B have varied little in 2013, increasing 36 pounds from either changes in the

manufacturing processes or more fidelity in the weight estimates from January through October 2013. Current estimates are within 202 pounds of the not-to-exceed weight of 32,577 pounds – the target weight of the aircraft in January 2015 to meet specification requirements and ORD mission performance requirements for vertical lift bring back. The small difference between the current weight estimate and the not-to-exceed weight allows for weight growth of 0.62 percent over the next year to meet technical specifications in January 2015.

- Managing weight growth with such small margins will continue to be a significant program challenge. Since the program will conduct the technical performance measurement of the aircraft in January 2015, well before the completion of SDD, continued weight growth through the balance of SDD will affect the ability of the F-35B to meet the STOVL mission performance KPP during IOT&E.
- Other F-35B discoveries included:
  - Wet runway testing, required to assess braking performance with a new brake control unit in both conventional and slow landing operations, has been delayed due to the inability to create the properly degraded friction conditions on the runways at the Patuxent River Naval Air Station, Maryland. The program plans to complete this testing in early CY14. Fielded F-35B aircraft at Eglin and at Yuma are operating under restricted landing conditions until the wet runway testing is complete.
  - Buffet and TRO continue to be a concern to achieving operational capability for all variants. The program made changes to the flight control laws to reduce buffet and TRO in the F-35B in CY13. No further changes to the control laws are being considered, as further changes will potentially adversely affect combat maneuverability or unacceptably increase accelerative loading on the aircraft's structure. The program plans to assess the operational effect of the remaining TRO and the effect of buffet on helmet-mounted display utility by conducting test missions with operational scenarios in late CY13 and early CY14.

F-35C Flight Sciences

Flight Test Activity with CF-1, CF-2, and CF-3 Test Aircraft
- F-35C flight sciences focused on:
  - Block 2B envelope expansion for weapons bay doors open and closed
  - Completing electromagnetic environmental effects testing to support shipboard operations
  - Surveying handling qualities in the transonic flight regimes
  - Regression testing of new air vehicle systems software
  - High angle-of-attack testing, which began in August
  - Carrier suitability testing in preparation for the first set of ship trials scheduled for mid-CY14. The program configured aircraft CF-3 with a modified and instrumented nose landing gear system to begin initial catapult testing in August 2013. The test team modified CF-3 with the new

arresting hook system and began on-aircraft testing with rolling engagements in late CY13.
- The test team completed three weapon safe-separation events by the end of October.
- The program modified one F-35C with new coatings on the horizontal tail, and similar to what was experienced in the F-35B and the F-35A, the coatings bubbled and peeled after experiencing high-temperature and high-airspeed conditions. These conditions involve extended use of the afterburner not expected to be representative of operational use, but which was necessary to achieve certain test points. The program plans to modify all three F-35C flight sciences aircraft with new tail coatings and temperature-sensing instrumentation to collect data to characterize conditions and determine what, if any, material solutions will be required. Non-instrumented test aircraft continue to operate with restrictions to the flight envelope and use of the afterburner.

Flight Sciences Assessment
- F-35C flight sciences test flights accomplished were ahead of the plan through the end of October, with 181 sorties completed compared to 171 planned.
- The test team lagged by 11 percent in completing the planned baseline test points through the end of October, accomplishing 1,032 points against the plan of 1,165 points. Progress through the Block 2B flight envelope lagged by 12 percent, as 947 of 1,080 points were accomplished. The test team was able to accomplish more test points in the Block 3F envelope than planned – completing 485 points, compared to 85 planned, pulling 400 points projected for completion in 2014 back into 2013.
- Weight management is important for meeting air vehicle performance requirements. The aircraft weight is computed monthly, and adjusted for known corrections from engineering estimates and production modifications.
  - The program added 139 pounds to the F-35C weight status in May 2013 to account for the redesigned arresting hook system. The latest weight status report from October 2013 showed the estimated weight of 34,593 pounds to be within 275 pounds (0.79 percent) of the projected maximum weight needed to meet technical performance requirements in January 2016.
  - This margin allows for 0.35 percent weight growth per year. The program will need to continue rigorous weight management through the end of SDD to avoid performance degradation and operational impacts.
- F-35C discoveries included:
  - Buffet and TRO continue to be a concern to achieving operational combat capability for all variants. Control laws have been changed to reduce buffet and TRO in the F-35A and F-35B with some success; however, both problems persist in regions of the flight envelope, and are most severe in the F-35C.
  - Characterization testing of buffet and TRO in the F-35C with the current control laws and without the use of

leading edge spoilers is ongoing. Unlike the other two variants, the program has the option to conduct flight testing with leading edge spoilers to reduce buffet and the onset of TRO with two of the F-35C flight test aircraft if trade-offs made in control laws are not sufficient to manage the negative impact of these effects.

Mission Systems

Flight Test Activity with AF-3, AF-6, AF-7, BF-17, BF-18, and CF-8 Test Aircraft and Software Development Progress

- Mission systems are developed and fielded in incremental blocks of capability.
  - Block 1. The program designated Block 1 for initial training capability and allocated two increments: Block 1A for Lot 2 (12 aircraft) and Block 1B for Lot 3 aircraft (17 aircraft). No combat capability is available in either Block 1 increment.
  - Block 2A. The program designated Block 2A for advanced training capability and designated this block for delivery of aircraft in production Lots 4 and 5. No combat capability is available in Block 2A.
  - Block 2B. The program designated Block 2B for initial, limited combat capability with internal weapons (AIM-120C, GBU-32/31, and GBU-12). This block is not associated with the delivery of any production aircraft. Block 2B software, once complete with development and certification, will be retrofitted onto earlier production aircraft.
  - Block 3i. The program designated Block 3i for delivery of aircraft in production Lots 6 through 8, as these aircraft will be built with an improved integrated core processor and other upgrades collectively known as "Technology Refresh 2", or TR2. No new capability beyond Block 2B is introduced in Block 3i.
  - Block 3F. The program designated Block 3F as the full SDD capability for production Lot 9 and later.
- The Edwards test site accepted the fifth F-35C test aircraft, designated as CF-8, in September 2013; it is a mission systems flight test aircraft.
- The six mission systems flight test aircraft assigned to the Edwards AFB test center flew 302 test sorties against a plan of 286 though October, exceeding the plan by 5.6 percent.
- However, the test team accomplished only 54 percent of the planned 2013 baseline mission systems test points from test plans for Blocks 1, 2A, and 2B by the end of October (955 baseline test points accomplished, 1,755 planned). The team also accomplished an additional 1,194 test points for regression testing of new revisions of Block 2A and 2B software and other testing the program found necessary to add to the test plans. The team also lagged in completing planned radar signature testing, completing 346 of 723 planned test points, or 48 percent, by the end of October.
- The program initiated a Block Review Board process in late 2012 to manage the increments of mission systems software development, monitor maturity of capability, and release to flight test.

Mission Systems Assessment

- Despite flying the mission systems test flights planned for CY13, the program did not make the planned progress in developing and testing mission systems capabilities. Software development, integration in the contractor labs, and delivery of mature capability to flight test continued to be behind schedule. Testing of Block 2A training capability (no planned combat capability) was completed in 2013. The first increment of Block 2B software, version 2BS1, was delivered to flight test in February 2013, four months later than indicated in the integrated master schedule.
- The program completed testing on the Block 2A software needed for delivery of the Lot 4 and Lot 5 production aircraft. This production version of software, designated 2AS3, was designed to provide enhanced training capabilities to the Integrated Training Center at Eglin AFB, Florida, and to the first operational units – the F-35B unit at Yuma Marine Corps Air Station, Arizona, and the F-35A unit at Nellis AFB, Nevada.
  - However, the teams at both test centers (Edwards and Patuxent River) determined the initial version of 2AS3 to be deficient in providing the necessary capabilities for unmonitored flight operations under night and instrument meteorological conditions (IMC). In order to finalize Block 2A capability so that it could eventually be certified in production aircraft for flight at night and in IMC, the program made adjustments to plans for the following increment, Block 2B, to accommodate the need for another, final version of Block 2A software, designated 2AS3.1. The test centers completed testing of Block 2AS3.1 in June; however, the certification to allow F-35A and F-35B production aircraft to fly at night or in IMC had not been released as of the time of this report.
  - Additionally, the test teams also noted Block 2A deficiencies in the aircraft sensor operations, particularly the Electro-Optical Targeting System (EOTS), aircraft communications capabilities, pilot electronic interfaces, and the aircraft Caution, Advisory, and Warning System. Although the software was intended to provide more mission systems capability, poor sensor performance and stability, excessive nuisance warnings, and disproportionate pilot workload required for workarounds and system resets made the software of limited utility for training. In any type of operational mission scenario, the performance of the software would be unacceptable.
  - The program delivered 10 F-35A aircraft to the U.S. Air Force, 12 F-35B aircraft to the U.S. Marine Corps, and 2 F-35C aircraft to the U.S. Navy from production Lot 4 through the end of October. These aircraft were delivered in the Block 2A configuration, but with less capability than defined by the production contract. Specifically, 22 of 47 (47 percent) of the capabilities defined in the production contract were not complete when the aircraft were delivered. The program began checkout and delivery of F-35A, F-35B, and F-35C aircraft from production Lot 5, and these aircraft were similarly delivered with less

than planned capabilities. Fifty percent (27 of 54) of the capabilities required by the contract were not complete when these aircraft were delivered to the Services.

- The initial Block 2B software increment began flight testing in February 2013. Though four months later than the 2012 integrated master schedule, this timing was in accordance with the expectations set by the program's new Block Review Board process, which was initiated in late 2012. As it was the initial Block 2B increment, no new capability was mature enough for verification. In October 2013, a new increment of Block 2B, intended to provide a significant increase in verifiable capability, including many fixes to previously identified deficiencies, began flight testing. Initial results with the new increment of Block 2B software indicate deficiencies still exist in fusion, radar, electronic warfare, navigation, EOTS, Distributed Aperture System (DAS), Helmet-Mounted Display System (HMDS), and datalink. These deficiencies block the ability of the test team to complete baseline Block 2B test points, including weapons integration. The program's plan is to gradually increase maturity of the software and reduce these obstacles to test progress over three more increments of software in CY14. The degree to which the maturity of the capability has improved and the test teams can verify performance against planned criteria will determine how long it will take to complete Block 2B development and flight test.
- The program began implementing plans for testing Block 3i capability, which will be used to deliver production aircraft in Lots 6 through 8, all of which will have an upgraded core processor and other mission systems processor improvements. The program plans Block 3i to include no new capability beyond Block 2B, as it is intended to only encompass rehosting of Block 2B capability on the new TR2 hardware.
  - One F-35A mission systems test aircraft was temporarily modified with the TR2 hardware in November 2013 to conduct risk reduction testing of an early version of 3i software. Testing was attempted on an F-35C test aircraft in October, which was temporarily modified with the TR2 hardware, but the software did not load properly and the ground testing could not be conducted.
  - One mission systems test aircraft of each variant will be modified in early CY14 to begin the start of flight testing of the 3i software.
  - All production aircraft from Lot 6 and beyond will have the TR2 hardware and will only be able to operate mission and vehicle systems software that is compatible with this hardware configuration.
- Shortfalls in the test resources required to test mission systems electronic warfare capabilities under operationally realistic conditions were identified by DOT&E in February 2012. The DoD programmed for an Electronic Warfare Infrastructure Improvement Program starting in FY13 to add both closed-loop and open-loop emitter resources for testing on the open-air ranges, to make at least

one government anechoic chamber capable of providing a representative threat environment for electronic warfare testing, and to upgrade the electronic warfare programming laboratory that will produce threat data files. However, progress has been slower than needed to assure these resources are available in time for Block 3 IOT&E in 2018. JSF IOT&E will not be adequate and will be delayed unless this test capability is available.

- Deficiencies in the HMDS added testing at both the Edwards and Patuxent River test sites in late CY12 and in CY13. The program dedicated 42 flights to investigating and addressing deficiencies in the HMDS. Seven aircraft from all three variants flew test missions from October 2012 through May 2013 to investigate jitter in the helmet display, night vision camera acuity, latency in the DAS projection, and light leakage onto the helmet display under low-light conditions. Although some progress has been achieved, results of these tests have been mixed.
  - Filters for reducing the effects of jitter have been helpful, but have introduced instability, or "swimming," of the projected symbology.
  - Night vision acuity was assessed as not acceptable with the current night vision camera, but may be improved with a new camera planned for inclusion in the next version of the helmet (referred to as the Gen III helmet) being considered by the program.
  - Latency with the DAS projection has improved from earlier versions of software, but has not yet been tested in operationally representative scenarios.
  - Light leakage onto the helmet display may be addressed with fine-tuning adjustments of the symbology brightness—a process pilots will have to accomplish as ambient and background levels of light change, adding to their workload.
  - Although not an objective of the dedicated testing, alignment and "double vision" problems have also been identified by pilots and were noted in the DOT&E report on the F-35A Ready for Training Operational Utility Evaluation.
  - Developmental testing has yet to be accomplished in the full operational flight envelope evaluating mission-related tasks, as the full combat flight envelope is not yet available. Use of the HMDS in the full envelope under operational conditions is needed to verify effectiveness of the HMDS. This might not occur until the Block 2B operational utility evaluation, currently planned for late 2015.
- Three factors create a significant challenge for completing developmental testing of Block 2B mission systems as planned before the end of October 2014: completing tests of prior blocks of mission systems capability, managing growth in testing, and constraints on test resources.
  - The test centers continue to accomplish a significant amount of test points originally designated for completion in prior blocks of mission systems capability. As of the

end of October, 34 percent of the baseline mission system test points accomplished in CY13 (326 of 955) were for capabilities in Block 1; 18 percent (168 of 955) were for capabilities in Block 2A, and 48 percent (461 of 955) were for Block 2B capabilities. The program intends to complete or delete the test points planned in these previous blocks by the time Block 2B capability completes development in late CY14. All program plans and schedules for the subsequent blocks of mission systems software (Block 3i and Block 3F) depend on this occurring so that the development laboratories and test venues can be converted and devoted to testing the Block 3 hardware configuration.

- The program continues to have significant growth in mission systems testing. Beyond the testing accomplished in late CY12 and CY13 for the helmet, additional testing has been required for regression testing of seven software loads delivered to flight test in CY13 through October, and for deficiencies in the EOTS, the radar, night flying qualities, and navigation systems. Dedicated testing added for the purpose of identifying problems with the helmet accounted for only 22 percent of the total mission systems growth in CY13 by the end of October; the remaining growth executed by the program exceeded the planning factors for added testing by over 40 percent. The program plans to complete Block 2B flight testing in October 2014; however, there is no margin for additional growth to meet that date. Projections based on the planned growth rate show that Block 2B developmental testing will complete in May 2015, approximately 7 months later than planned. Projections for completing Block 2B flight testing using the historical rate of continued growth (excluding the growth associated with the HMDS) show that Block 2B developmental testing will complete about 13 months later, in November 2015, and delay the associated fleet release to July of 2016.

- Mission systems SDD flight test aircraft available to support Block 2B developmental testing will be reduced in CY14, as the program will need to modify aircraft with the TR2 processors to achieve the Block 3i configuration. Aircraft from production Lot 6, which are scheduled to be delivered in mid-CY14, cannot be operated with Block 2B software; they must have certified Block 3i software. The program plans to modify one mission systems aircraft of each variant to begin flight testing of the first increment of Block 3i software in early CY14. The reduction of mission systems aircraft to support Block 2B developmental testing, created by the need to test software to support the production and delivery of Lot 6 and later aircraft, will add to the challenges of completing Block 2B development on schedule.

- Mission systems discoveries included:
  - Although improving, stability of the mission systems software continues to fall short of objectives. The program tracks mission systems software stability by analyzing the number of anomalies observed as a function of flight time. The program objective for time between resets for the integrated core processor and the Communications/Navigation/Identification Friend or Foe suite is a minimum of 15 hours between reset events. October reports for the latest Block 2B mission systems software increment in flight test show a rate of 11.4 hours between anomalies, based on 79.5 hours of flight test. Subsystems, such as the radar, EOTS, DAS, and the navigation solution often require component resets as well, but these are not tracked in the stability metric.
  - The EOTS fails to meet target recognition ranges, exhibits track instability in portions of its field-of-view, and has large line-of-sight angle and azimuth errors when computing target locations. These deficiencies are being investigated and addressed by the program with software fixes.
  - The program continues to monitor loading of the aircraft core processors in the laboratories as more functionality is added in software increments. Projections of the loads expected on all processors for the Block 3 capabilities estimate that three processors, which support landing systems, weapons employment, multi-aircraft datalinks, and earth spatial modeling, will be tasked between 160 and 170 percent of capacity. The program intends to shift the distribution of processing loads with each incremental build of mission systems software; however, margin is limited and the efficiencies gained by the changes need to be assessed under actual, sensor-stressing, flight conditions.
  - The DAS has displayed a high false alarm rate for missile detections during ownship and formation flare testing. The inability of the DAS to distinguish between flares and threat missiles makes the warning system ineffective and reduces pilot situational awareness.
  - The onboard navigation solution – referred to as the ownship kinematic model – has shown excessive position and velocity errors when not receiving updates from the GPS satellite constellation. These errors prevent accurate targeting solutions for weapons employment in a GPS-denied environment. The program is addressing these errors in the next iteration of software and further flight testing will be required.
  - The radar mapping function does not provide adequate target location accuracy.

Weapons Integration

- Weapons integration involves flight sciences testing, mission systems testing, and ground crew support. Testing includes measuring the environment around the weapon during carriage (internal and external), handling characteristics of the aircraft, safe-separation of the weapon from the aircraft, communications between the aircraft sensors and the weapons, and weapons delivery accuracy events. The program has identified lethality, the product of weapons integration test and evaluation, as the critical path to completing development of Block 2B and Block 3F. The Block 2B weapons are the GBU-12 laser-guided bomb, the GBU-31/32 JDAM, and the AIM-120 air-to-air missile. The Block 3F weapons add Small Diameter Bomb Increment I (SDB-I), AIM-9X air-to-air missile, Joint Standoff Weapon, gun (internal for F-35A and external gun pod for F-35B and F-35C), and the United Kingdom's Paveway IV bomb.

- As of the end of October, weapons integration was near the planned progress scheduled for the year on the F-35A. The test teams had completed 567 of 589 planned environmental test points and all 19 planned weapons separation events. Progress on the other variants, however, was behind the plan. On the F-35B, the team had completed 285 of the 455 planned environmental test points and 12 of the 24 planned separation events. On the F-35C, the team began environmental testing late in the year and had completed 176 of 181 planned test points but only 2 of 10 planned separation events.

- Progress in testing of mission systems capability to enable end-to-end weapon delivery events was behind schedule for all Block 2B weapons. Weapons integration has been slowed by discoveries of deficiencies requiring software fixes and additional testing.
  - Problems with the lanyard on the laser-guided bomb required a new lanyard and routing procedure
  - Inaccuracies in the data transfer of position and velocity from the aircraft to the JDAM, which spatially align the bomb with the target, required a fix in the mission systems software
  - Problems involving integration of the AIM-120 medium-range missile have been difficult to replicate in lab and ground testing
  - Poor target track quality displayed to the pilot from the radar, or from fusion of the aircraft sensors, prevented targeting solutions for simulated weapons engagements
  - Poor performance of the EOTS in image quality, tracking stability, and targeting accuracy required software fixes to allow weapons integration testing of the air-to-ground munitions to proceed
  - Erroneous target coordinates were derived from the synthetic aperture radar mapping function

- The integrated test team continued to rework weapons integration scheduling in 2013 to account for discoveries of deficiencies and the slower than expected delivery of capability needed to conduct weapons delivery accuracy (WDA) events. The team conducted the first WDA test event with a laser-guided bomb on October 29, followed two days later by the first launch of the AIM-120 air-to-air missile. The second launch of an AIM-120 missile occurred on November 15. Data analyses of the missile launches was ongoing at the time of this report. The team accomplished the first WDA test event with a JDAM bomb (GBU-32) on December 6; data analysis was ongoing at the time of this report. These early WDA events have included non-operationally relevant workarounds to mission systems

deficiencies that will not be tolerable in operational testing or combat employment. Completion of all Block 2B weapons testing by the end of October 2014 is dependent on:
- The ability of the test team to accomplish a successful weapons-related test mission at a consistently high rate
- The Block 2B version of mission systems software delivered in October 2013 adequately correcting deficiencies and permitting WDA events to proceed in an operationally relevant manner
- Reliable instrumentation and priority from range support assets
- Maintaining the test aircraft used for weapons testing in the Block 2B configuration while the program manages the requirement to start testing mission systems aircraft in the Block 3i configuration

- Current program schedules indicate weapons integration testing to be complete by the end of October 2014 and August 2016 for Blocks 2B and 3F, respectively. To meet the schedule for Block 2B, the test team planned to have completed 8 of 15 total Block 2B WDA events by the beginning of December; however, only 4 have been accomplished. WDA events beyond these first four have been blocked from completion due to lack of adequate mission systems performance in radar, fusion, and EOTS. Corrections to the known deficiencies and fix verification are planned to be delivered in the 2BS4.2 and 2BS5 versions of software, the first of which is scheduled to begin weapons flight testing in March 2014. The result of this blocking of subsequent WDA events is a 4- to 6-month delay in the completion of Block 2B weapons integration, which will likely be done between February and April 2015. Detailed planning of the Block 3F weapons integration schedule to complete in August 2016 is under development. However, given historical performance and reasonable planning factors, it is more likely that the final Block 3F weapons events will not be completed within the current SDD schedule.

Static Structural and Durability Testing

- Durability testing and analysis on the ground test articles of all three variants continued in 2013; progress is measured in aircraft lifetimes. An aircraft lifetime is defined as 8,000 Equivalent Flight Hours (EFH), which is a composite of time under different test conditions (i.e., maneuver and buffet for durability testing). In accordance with the SDD contract, all three variants will complete two full lifetimes, or 16,000 EFH of durability testing. The completion dates for the second aircraft lifetimes are late 2014 for the F-35B and early 2015 for the F-35A and F-35C. The program made plans in 2013 to add a third lifetime of durability testing on the test articles of all three variants.

- The F-35A ground test article, AJ-1, completed the first aircraft lifetime in August 2012, as planned. For most of 2013, AJ-1 underwent detailed inspections and repairs on cracks revealed after the first lifetime of testing, including repairs to the wing forward root rib and to a bulkhead stiffener. The second lifetime of durability testing is planned to begin in December 2013.

- F-35B durability testing on BH-1 completed the first lifetime of 8,000 EFH on February 9, 2013, then underwent detailed inspection and repairs prior to starting the second lifetime of testing on July 22. The program completed the first block of 1,000 EFH (9,000 EFH total) on August 19, approximately 1 month ahead of schedule. Further testing was halted in September when cracks were discovered in two of the bulkheads, requiring repair.

- The F-35C fatigue test article restarted testing on January 9, 2013, after previously completing 4,000 hours of testing and associated inspections. It completed 8,000 EFH of testing, or the first lifetime, on September 28. Testing is behind schedule, as cracks discovered in the floor of the avionics bay in February caused a two-month pause while interim repairs were completed. Cracks discovered in fuselage station 402 and the surrounding structure caused a stop test after 7,620 EFH of testing to complete repairs. These cracks were not predicted by prior analysis. Detailed inspections from the first lifetime were ongoing as of this report.

- Component durability testing for two lifetimes of the vertical tails was completed for the F-35A and F-35B during 2012. Vertical tail testing started in August 2012 for the F-35C and completed 12,901 EFH as of the end of October 2013. Component testing of the horizontal tail for the F-35A and F-35C began third-lifetime testing, completing 23,000 EFH and 21,000 EFH, respectively, as of the end of August.

- The redesigned F-35B auxiliary air inlet doors, required for STOVL operations, are undergoing ground tests on the F-35B static loads test article (BG-1). Static load testing was completed late in CY12 and durability testing had completed just over 3,000 cycles (approximately 8 percent) of the planned testing as of the end of August. Modifications of the auxiliary air inlet doors on production aircraft have already begun.

- Discoveries from durability testing included significant findings in both the F-35A and F-35B ground test articles.
  - Discoveries this year on the F-35A test article include cracks in the engine thrust mount shear webs (designed to carry some of the fore and aft engine load) on both sides of the aircraft, and a crack in the frame of the web stiffener located at fuselage station 402. The program has redesigned the thrust mounts for production cut-in with Lot 6, and retrofits to be completed on earlier aircraft during depot modification periods. Root cause, corrective action, and modification plans for the frame crack are to be determined.
  - In the F-35B, the program halted testing in December 2012 after multiple cracks were found in a bulkhead (FS472) flange on the underside of the fuselage during the 7,000-hour inspection. Root cause analysis, correlation to previous model predictions, and corrective action planning are ongoing.
    - Discoveries during detailed inspections following the first lifetime of testing include cracks on the left and right hand sides of the wing aft spar lower flanges and

cracking in the frame of the jack point stiffener, a portion of the support frame outboard of the main fuselage above the main landing gear designed to support load bearing of the aircraft during jacking operations. Redesign, modification, and retrofit plans for these discoveries have not yet been determined by the program. As of August 5, 2013, two redesigns of the part were being evaluated for potential replacement.

- During its 8,000-hour detailed inspection period between February and July, cracks were found on both the right and left rear spar lower flanges near bulkhead FS556. This particular spar was already on the list of limited life parts, but not for the location of concern.

- Also during its 8,000-hour inspections, cracks were found in the lower arch of the FS496 bulkhead, but were below limits which would cause a break in planned testing, which restarted at the end of July. At the 9,000-hour inspection in September, the cracks had grown, but were not deemed sufficient to stop testing, but required increased inspection intervals. The cracks continued to grow during subsequent testing, until at 9,056 EFH, at the end of September, the bulkhead severed and transferred loads which caused cracking in the adjacent FS518 bulkhead. Analysis and corrective action were ongoing at the time of this report.

- All of these discoveries will require mitigation plans and may include redesigning parts and additional weight. Also, the repairs to the jack point stiffeners – accomplished after the first lifetime of testing – were not adequate, requiring the program to design a new repair concept.

- Discoveries in the F-35C test article include cracks in the floor of the avionics bay and, similar to the F-35B, cracking in the frame of the jack point stiffener. Cracks were also found in the bay floor of the power distribution center; repair, retrofit, and production impacts are to be determined.

Modeling and Simulation

Verification Simulation (VSim)

- VSim is a man-in-the-loop, mission software-in-the-loop simulation developed to meet the operational test agencies' requirements for the Block 2B operational utility evaluation and Block 3F IOT&E.

- The program is now at significant risk of failing to (1) mature the VSim and (2) adequately verify and validate that it will faithfully represent the performance of the F-35 in the mission scenarios for which the simulation is to be used in operational testing. Key concerns are:
  - VSim development, and verification and validation activities may not be completed in time to support the Block 2B operational utility evaluation, beginning in late CY15. In particular, long lead items such as threat mission data files are at risk of being delivered too late for integration into VSim in time to support the planned Block 2B operational utility evaluation timeline.

Additionally, the current VSim schedule has validation and accreditation documentation production activities scheduled until September 2015, months late to support the initial accreditation report required by the Operational Test Readiness Review for the Block 2B operational utility evaluation, scheduled for May 2015.

- The current VSim validation plan does not provide the detail or rigor needed to be able to anticipate accreditation of VSim for use in mission-level evaluation in operational testing. Shortfalls identified include: lack of detail in validation plans for VSim component models; lack of a clear path from component model validation to F-35 system validation to mission-level validation; absence of planned validation for government-furnished threat and weapons models that require significant additional validation after the modifications made to them during integration into VSim; and lack of a plan for structured regression testing after model modifications have been made. As of November 2013, the JSF Operational Test Team, the JSF Program Office, and Lockheed Martin are in the midst of a series of intensive VSim validation meetings aimed at overcoming these shortfalls.
- VSim may not adequately replicate the installed system performance (i.e., the performance of all F-35 systems and subsystems as installed in the aircraft) in the mission scenarios for which the simulation is planned to be used in the Block 2B operational utility evaluation. There may not be adequate validation data to support accreditation of the simulation for operational testing.
- No dedicated testing is planned by the program to validate F-35 installed performance in the VSim. The program currently expects validation data to come from planned developmental mission systems and weapons integration testing. However, developmental testing seeks only to acquire verification of contract specification criteria, and does not span the set of conditions over which mission effectiveness will be assessed using VSim in both developmental and operational testing. This creates a significant gap for the program in being able to validate VSim for both developmental and operational testing.

• In addition to the risks cited above, DOT&E has highlighted shortfalls in the test resources needed to gather key elements of data required for validation of the VSim for IOT&E, in particular for electronic warfare performance in the presence of advanced threats. These shortfalls are a function of limitations in the test assets currently available to represent threat systems. DOT&E has made formal recommendations to address the shortfalls and is pursuing solutions to make the assets available in time to prepare for IOT&E in a realistic threat environment.

• The JSF Program Office and Lockheed Martin have begun to try to address these concerns. Important recent activities have included technical interchange meetings with threat model developers in the intelligence community to address the modeling of electronic attack capabilities, a series of intensive validation planning meetings currently underway to provide detailed validation data requirements, and a summer 2013 VSim risk reduction event using the simulation in an F-35 Block 2A configuration.

Other Models and Corporate Labs Activity

• At the beginning of 2013, the Program Office had accredited 7 of the 25 models and simulations currently planned to support verification of the F-35. No additional models and simulations planned to support verification of F-35 requirements were accredited in 2013; so, the total number accredited remains at seven.

• As of the end of 2012, the program had planned to accredit six models and simulations intended for use in the requirements verification plan in 2013. Of the 18 remaining models and simulations listed in Program Office documentation as requiring accreditation for use in verification, the program characterizes 12 as on-track for accreditation. The progress of the remaining six is characterized as either off-track with mitigation efforts in place or as on-track but with significant execution risk.

Training System

• In late 2012, the program completed a Ready For Training Operational Utility Evaluation (OUE) to support the Air Force's Air Education and Training Command's decision to begin student training at Eglin AFB, Florida. The OUE evaluated the capability of both the F-35A air vehicle and the training system to train an experienced initial cadre of pilots in the equivalent of the familiarization phase of a fighter aircraft transition syllabus. It also evaluated the ability of the F-35A maintenance and Autonomic Logistics Information System (ALIS) to sustain a sortie generation rate for the Block 1A syllabus.

• Restrictions on the aircraft operating limits prevented instruction in most high performance maneuvering and flight through instrument meteorological conditions (i.e., clouds). However, pilots were adequately trained in the basic operation of the aircraft. Mission systems were still immature, but generally unnecessary for this phase of training since no combat training could be performed. Even at this reduced level of activity, the radar, the HMDS, and the cockpit interfaces caused increased workload or had deficiencies. Aircraft availability was low during the OUE, but was adequate to meet the training sortie requirements with extensive workarounds.

• Pilot training classes continued throughout 2013. Although aircraft availability and reliability at the training center remains below expectations, the shortened syllabus allowed pilot production to remain at planned levels. Eglin originally planned to produce 68 pilots during the 2013 period of performance, but the Services reduced their need to 66 pilots. All students completed planned training (of the reduced syllabus) on schedule.

- There are currently two configurations of aircraft at the training center, Block 1B and 2A. Six Lot 4 (Block 2A) aircraft were delivered in 2013 and several Lot 5 aircraft are in various stages of delivery. The first two F-35C aircraft were delivered to Eglin AFB in June. Pilot training using the syllabus for the Block 2A configuration starts in early 2014 after a small group rehearsal.
- The training center continued to conduct maintenance training for experienced maintenance personnel for both the F-35A and F-35B during 2013. As of the end of September, 978 personnel had completed training in one or more of the maintenance courses to support fielded maintenance operations.

Live Fire Test and Evaluation

F135 Engine

F135 engine vulnerability testing consisted of two test series: (1) fuel ingestion tests to examine the vulnerability of the F135 engine caused by fuel leakage from ballistically damaged fuel tanks adjacent to the engine inlets, and (2) ballistic tests to determine the damage tolerance of engine components, including fluid-filled components, sensors, actuators, and rotating components.

- The fuel ingestion tests demonstrated the engine can tolerate a range of inlet fuel flows. These fuel flow rates simulated quantities representative of missile fragment-induced damage to fuel tanks adjacent to the engine. System-level ballistic test events planned for FY15, using a structural F-35C test article with an operating engine, will quantify the exact relationship of the simulated leak rates to those expected in an actual threat encounter. Further analysis will assess the vulnerability to multiple fragment impacts, which are probable in missile encounters.
- The fuel ingestion tests did not simulate engagements by ground-based or aircraft gun systems that are possible during low-altitude close-air support missions and within-visual-range air-to-air combat. A Concept Demonstrator Aircraft engine test in 2005 showed the engine could not tolerate fuel ingestion events representative of such conditions (i.e., low-altitude, high-speed, high-engine thrust, and higher leak rates). The program made no design changes in response to those earlier test results and this vulnerability remains in the final production engine design. A ballistic liner in the fuel tank could mitigate this vulnerability, but the program removed this feature during its weight-reduction efforts, saving 48 pounds.
- Tests using single missile fragments showed that the F135 rotating components were tolerant to these threats, with little or no effect on engine performance or component survival. However, three of four tests against fuel-filled external components resulted in massive fuel leaks, and one produced a sustained fire. The F-35C system-level tests in FY15 will evaluate whether installation effects, resulting in leaked fuel interacting with the engine exhaust, would increase the risk of fire. Engine vulnerability to high-explosive incendiary (HEI) and armor-piercing incendiary (API) threats was not confirmed in this test series since historical data on similar engines already demonstrated that these threats can penetrate the engine core and create cascading damage resulting in engine failure and fires.

F-35B Lift System

- Ballistic tests on an F-35B STOVL propulsion system showed that single fragment damage to the lift fan did not degrade propulsion system performance. Analyses showed that fragment-induced damage could result in the release of more than 25 percent of a single lift fan blade, resulting in a catastrophic STOVL system failure. In order to preserve the test article for the remainder of the series, these engagement conditions were not tested. More severe threats, encountered at low-altitude or in air-to-air gun engagements, will likely cause catastrophic damage.
- Ballistic tests of the lift fan shaft demonstrated that the design changes from the earlier Concept Demonstration Aircraft article improved its survivability against all threats, including the more severe API threat.
- The F-35 has no sensors to warn the pilot of lift fan damage prior to conversion to STOVL flight upon return for landing. Conversion to STOVL flight puts high loads on the quickly accelerating system components that can result in catastrophic failure before the pilot can react and return the aircraft to wing-borne flight, or can create uncontained damage that cascades into other critical system failures. Prognostics and Health Management sensors that monitor component health and system degradation for maintenance purposes, could provide some warning, but the relevant software and hardware would have to be improved to provide reliable information to the pilot to support critical survivability decisions.

On-Board Inert Gas Generation System (OBIGGS)

- An OBIGGS/lightning protection Critical Design Review in February 2013 reviewed a system design capable of providing fuel tank inerting that would prevent fuel tank ullage explosion due to ballistic threat encounters or lightning strikes. The program is currently planning the F-35B fuel system simulator testing and ground tests on all three variants. Tests will include a spectrum of mission profiles, including high descent-rate dives to evaluate the improved OBIGGS ability to provide fuel tank inerting without compromising fuel tank and wing structure integrity.
- In-flight inerting does not protect the aircraft against damage to the airframe resulting from lightning-induced currents. Most line-replaceable units (e.g., actuators and components of the electrical power system) have passed lightning tolerance qualification testing, but the existing F-35 airframe fasteners, selected to satisfy weight reduction criteria, are not lightning tolerant. The program still needs to complete lightning tolerance qualification testing for remaining components and current injection tests, before lifting current restrictions preventing aircraft operations within 25 miles of known lightning.

Polyalphaolefin (PAO) Shut-Off Valve

- A live fire test in 2012 demonstrated crew and aircraft vulnerabilities to avionics coolant (PAO) system fires. The threat ruptured the PAO pressure line in the area just below the cockpit, causing a sustained PAO based fire with a leak rate of 2.2 gallons per minute (gpm). These results showed that a PAO shut-off valve that could detect and react to a 2 gpm, low leak rate could mitigate this vulnerability. Designing a system with this criterion poses some technical challenges, given a potential for excessive false alarms at these detection rates.
- DOT&E repeatedly recommended redesigning and reinstalling a PAO shut-off valve after the program decided on removal for weight reduction. The program has been reconsidering the reinstatement of the PAO shut-off valve and has tasked Lockheed Martin to develop a technical solution to meet the criteria demonstrated in live fire tests. The program has not provided any updates on the operational feasibility and effectiveness of the design, or an official decision to reinstate this vulnerability reduction feature.

Fueldraulic Fuses

- The fueldraulic system is a fuel-based hydraulic system used to control the F-35B engine exhaust nozzle. It introduces a significant amount of fuel plumbing to the aft end of the engine and, consequently, an increased potential for fire. A live fire test in 2012 demonstrated the fueldraulics system is vulnerable to missile fragments, resulting in potential fire and loss of aircraft. Engine ballistic tests in FY13 also showed that the fueldraulics system is vulnerable and that a shut-off for a damaged system could mitigate much of the vulnerability.
- A fueldraulic shut-off feature could also provide safety-related protection. In 2013, prior to a routine flight test, testers discovered an F-35B fueldraulics line failure due to an improperly manufactured hose that could have led to an engine nacelle fire. An effective fueldraulic shut-off would prevent such an outcome.

Electrical System

- The F-35 includes several technologies used for the first time in a fighter aircraft that represent advancement of the more electric aircraft topology. The advances also provide a potential source of unique F-35 vulnerabilities.
- All flight control electronic units and the electrical power system electrical distribution units have two voltage levels (270 and 28 volts DC) in internal circuits. An in-flight incident in 2007, electrical arcing tests in 2009, and the flight-critical system-level test events in 2012 showed that the vulnerability of the F-35 electrical power system requires further analyses to address the likelihood and significance of ballistically induced arcing between the 270-volt and 28-volt electrical systems.
- Lockheed Martin also confirmed that all three F-35 variants include up to 28 wire harnesses that contain both 28- and 270-volt wires, but the contractor is still working on providing the comprehensive extent and locations of these harness runs. Lockheed Martin should conduct a vulnerability analysis as soon as possible to determine the likelihood of ballistically- or lightning-induced arcing from the 270-volt on a 28-volt system and to determine whether the resulting damage effects would be catastrophic to the airplane. DOT&E will review these analyses to provide a comprehensive assessment of the F-35 vulnerability to ballistic damage to the electrical power system.

Chemical/Biological Vulnerability

The program continues to make progress in the development of the decontamination system in preparation for the full-up system-level test planned for FY17.

- The F-35 Chemical Biological Warfare Survivability Integrated Product Team oversaw design and construction of a full-scale shelter liner and associated portable process containment shelter for chemical and biological decontamination operations. The contractor will set up the initial demonstration of shelter and liner for a form, fit, and function demonstration in 1QFY14 in conjunction with the Tactical, Cargo, and Rotary-Wing Aircraft Decontamination device. A full-scale setup at Edwards AFB in FY14 will demonstrate performance of the integrated liner, shelter, and decontamination system in preparation for the FY17 full-up system-level test of the apparatus with F-35 test article BF-4.
- The Integrated Product Team is coordinating closely with the Joint Program Executive Office for Chemical and Biological Defense in developing the F-35 Joint Strike Fighter variant of the Joint Service Aircrew Mask. The mask, scheduled to undergo a Critical Design Review in 1QFY14, has high-schedule risk because its development is contingent on mask integration with the F-35 HMDS. The Mask Program Manager expects an LRIP version of the mask to be available in 3QFY14 in preparation for Mask/HMDS flight qualification in 1QFY15.

Gun Ammunition Lethality and Vulnerability

- The F-35 program, the Air Force, Navy, Marines, and their international partners are conducting lethality live fire testing and evaluation of three different 25 mm gun ammunition types.
  - PGU-48 frangible tungsten armor piercing design for the F-35A
  - PGU-32 semi-armor piercing HEI ammunition for the F-35B and F-35C
  - PGU-47 armor-piercing explosive ammunition for the partner F-35A variant and, depending on the overall cost and final lethality and reliability assessment results, possibly for the U.S. F-35B and F-35C variants
- Each ammunition is specialized against different target sets particular to each Service, including personnel, small boats, ground structures, trucks, light armor, and fixed-/rotary-wing aircraft.

- Fracture characterization tests of the PGU-48 showed the tungsten to be much more frangible than other tungsten materials tested previously, which should increase predicted damage against targets employing widely-spaced materials. Characterization of all three ammunitions will continue in FY14 with terminal ballistics tests against multi-plate structures (representing vehicle materials) as well as building wall materials. FY15 tests will include ground-based and flight testing against representative targets.
- The program assessed the vulnerability of the F-35 aircraft to ballistic threats while carrying these ammunitions in FY13. Ballistic tests against a single F-35 ammunition type (PGU-32) showed that propellant explosive reaction was highly unlikely, while a propellant fire was probable. No propellant fire generated by ballistic impact triggered a propellant explosion. There was no evidence of sympathetic reactions in multiple round tests.

Issues Affecting Operational Suitability

Overall suitability performance continues to be immature, and relies heavily on contractor support and workarounds unacceptable for combat operations. Aircraft availability and measures of reliability and maintainability are all below program target values for the current stage of development.

F-35 Fleet Availability

- Average F-35 availability rates for operational units are below established threshold values. (Availability is not a meaningful metric for aircraft dedicated to test, and thus SDD aircraft are not included in this section.)
  - The program established an availability threshold rate of 50 percent and an objective rate of 75 percent to track fleet performance for Performance Based Logistics agreements.
  - Aircraft availability rates by operating location from November 2012 through October 2013 are summarized in the following table. The first column indicates the average availability achieved for the whole period, while the maximum and minimum columns represent the range of monthly availabilities reported over the period.

F-35 AVAILABILITY FROM NOVEMBER 2012 THROUGH OCTOBER 2013*			
Operational Site	Average	Maximum	Minimum
Whole Fleet	37%	46%	26%
Eglin F-35A	38%	51%	24%
Eglin F-35B	39%	54%	22%
Eglin F-35C **	32%	61%	13%
Yuma F-35B	29%	45%	6%
Edwards F-35A	29%	41%	14%
Nellis F-35A	37%	63%	14%
		* Data do not include SDD aircraft	
		** Eglin F-35C data began in August 2013	

- Overall fleet availability has averaged 37 percent and showed a gradual decline in the latter half of the period reported in the table, with the last five months of the period all below

the average for the year. Late in the reporting period, the program began increasing the number of aircraft undergoing modifications and depot-level repairs, which contributed to the decline in fleet availability. While some operating sites did achieve threshold availability for a month or more, overall fleet availability never reached the threshold of 50 percent and was as low as 26 percent in February.

- Unavailable aircraft are considered Not Mission Capable (NMC) because they are undergoing maintenance (NMC-M) for systems necessary for safe flight or are awaiting parts from supply (NMC-S).
  - From November 2012 through August 2013, the NMC-M rate averaged 35 percent and was generally stable, but rose afterward and peaked at 47 percent in October. This observed NMC-M rate is well above the target rate of 6 percent established by the program for Performance Based Logistics evaluation.
    - A significant portion of the aircraft down time has been the result of field maintenance organizations waiting for technical dispositions or guidance from the contractor on how to address a maintenance issue that has grounded an aircraft. These Action Requests (ARs) are a result of incomplete or inadequate technical data in the field, and waiting for their resolution accounts for 25 to 30 percent of the aircraft downtime. Recent trends have shown an increasing number of ARs per aircraft each month. Reducing the rate of ARs, or decreasing the response time to the ARs, should improve NMC-M rates.
    - The requirement for modifications will continue to increase on the fleet and will likely adversely affect NMC-M rates for the next two years. Analysis of current modification plans show that up to 13 percent of the fielded fleet would be unavailable due to depot work alone in the late 2014 timeframe.
  - Over the same period, the NMC-S rate averaged 27 percent, peaking at just over 30 percent in July 2013 and then gradually declining. The target value established by the Program Office is an NMC-S rate of 20 percent or less. According to the Program Office, lower than expected performance in NMC-S rates has been due to late contracting of the necessary spares for recent production lots. They expect that improved contracting performance and increasing maturity of the supply system will result in improved parts support by late 2014.

F-35 Fleet Reliability

- The F-35 program uses reliability growth curves that project expected reliability for each variant throughout the development period based on accumulated flight hours.
  - These growth curves are established to compare observed reliability with a target to meet the Mean Flight Hours Between Critical Failure (MFHBCF) threshold requirement by 75,000 flight hours for the F-35A and F-35B, and by 50,000 flight hours for the F-35C.

- Currently, none of the variants are achieving their predicted reliability based on flight hours accumulated as of the end of August 2013, as shown in the following table.

F-35 RELIABILITY AS OF AUGUST 31, 2013 – MFHBCF, HOURS							
	Requirement		Current Values				Observed MFHBCF as of May 2012
Variant	Threshold MFHBCF	Threshold Flight Hour Target	Observed MFHBCF	Current Total Flight Hours	Objective MFHBCF from Growth Curve	Observed as % of Objective Value	
F-35A	20	75,000	4.5	4,204	13.5	33%	5.9
F-35B	12	75,000	3.0	3,286	7.7	39%	4.2
F-35C	14	50,000	2.7	903	9.0	30%	6.7

- Though month-to-month reliability rates vary significantly, in part due to the small fleet size, the F-35B showed slight improvement over the reporting period, while F-35A reliability appears to be relatively flat. The program has fielded too few F-35C aircraft to assess reliability trends.
- Statistical analysis of the 90-day rolling averages for Mean Flight Hours Between Critical Failure – Design Controllable ($MFHBCF_{DC}$) through the end of July 2013 show flat trend lines for the F-35A and F-35B with most data points below the threshold growth curve, meaning the observed reliability is not within the desired envelope for design controllable failures. Design controllable failures are those that can be attributed to deficiencies in component design, but considered by the Program Office to be fixable by design modification.
  - While some design improvements will be incorporated in production of the Lot 5 aircraft, most of the remaining planned improvements are being incorporated in Lots 6 and 7. The next opportunity to expect improvement in the fleet reliability performance is likely to be in 2015. However, some design improvements planned to be cut-in with these production lots are for structural fatigue life and increased mission capability which will not necessarily improve reliability.
  - Through November 2013, all F-35 test and production aircraft combined had achieved 11,500 total flight hours, 6 percent of the flight hour total (200,000 hours) at which the ORD reliability goal is to be achieved. However, the design is becoming more stable and opportunities for reliability growth are decreasing. While the relatively low number of flight hours shows there is still time for program reliability to improve, this is not likely to occur without a focused, aggressive, and well-resourced effort.
- A number of components have demonstrated reliability much lower than predicted by engineering analysis, which has driven down the overall system reliability. High driver components affecting low availability and reliability include the following, grouped by components common to all variants as well as by components failing more frequently on a particular variant or completely unique to it, as shown in the following table.

HIGH DRIVER COMPONENTS AFFECTING LOW AVAILABILITY & RELIABILITY		
	Specific to Variant	Common to All Variants
F-35A	• Data transfer cartridge • Position/strobe light lens assembly	• 270 Volt Direct Current battery • Fiber channel switch • Avionics processor • Power and thermal management system • Landing gear and tire assembly • Display management computer/helmet • On-Board Oxygen Generating System • Crew escape and safety system • 80kW Inverter/Converter/Controller
F-35B	• Upper lift fan door actuator • Main landing gear wheel/tire assembly	
F-35C	• Ejection seat portion assembly • Data security module	

Maintainability
- The amount of time required to repair failures for all variants exceeds that required for mature aircraft, and has increased over the past year. The table below compares the Mean Corrective Maintenance Time for Critical Failure (MCMTCF) and Mean Time To Repair (MTTR) for all unscheduled maintenance for each variant as of August 31, 2013, to the threshold requirement from the ORD and the same value reported in the FY12 Annual Report.

F-35 MAINTAINABILITY AS OF AUGUST 31, 2013 - MCMTCF (HOURS)				
Variant	Threshold	Observed	% of Threshold	FY12 Annual Report
F-35A	4.0	12.1	303%	9.3
F-35B	4.5	15.5	344%	8.0
F-35C	4.0	9.6	241%	6.6

F-35 MAINTAINABILITY AS OF AUGUST 31, 2013 - MTTR (UNSCHEDULED)				
Variant	Threshold	Observed	% of Threshold	FY12 Annual Report
F-35A	2.5	9.2	366%	4.2
F-35B	3.0	8.9	294%	5.3
F-35C	2.5	7.7	307%	4.0

- Maintenance times reported by the Program Office have increased (worsened) compared to those reported a year ago.
  - The causes of this increase are not clear from the available data, which are derived from a fleet that has only early mission systems functionality, but has grown to include three new operating locations this year. It is too early to determine if the increase in maintenance times is from immaturity of sustainment operations in the field (i.e., incomplete technical data and low experience of newly-trained maintenance personnel) or from underlying maintainability and aircraft design issues, such as poor component reliability and maintenance actions requiring excessive time to complete.
  - Cure time to restore low-observable (LO) characteristics following maintenance behind panels not designed for frequent access might be a factor in the increased maintenance time, but the Program Office has not tracked

LO maintenance times separately. The Program Office should include LO and non-LO repair times in their monthly performance metrics to help understand the root cause of these increases and take corrective actions. Further, LO repair should be broken down into repair times for inherent LO failures, and LO repairs required to facilitate other maintenance. The proportion of all LO repairs that are required to facilitate other maintenance should be reported.

Autonomic Logistics Information System (ALIS)

- The Program Office continues to develop and field ALIS in incremental capabilities similar to the mission systems capability in the air vehicle. Overall, the ALIS is immature and behind schedule, which adversely affects maintainability and sortie generation. Shortfalls in functionality and data quality integrity require workarounds and manual intervention.
- ALIS version 1.0.3, required for the Services to accept production Lot 4 aircraft at Eglin AFB, Florida, Nellis AFB, Nevada, and Yuma Marine Corps Air Station, Arizona, underwent initial testing at the Edwards test center in late 2012 and began fielding in early 2013.
  - During initial testing in 2012, the Edwards test team found shortcomings in the systems integration of ALIS applications and a lack of maturity in handling data elements. The team identified four critical (Category I) deficiencies, which required correction before fielding, and 54 severe (Category II) deficiencies, which required significant workarounds.
  - The contractor developed an updated version of the ALIS 1.0.3 software to address some of the deficiencies identified during initial testing and the Edwards test team retested the software in December 2012. The program subsequently started fielding this version of ALIS 1.0.3 in early 2013.
  - The Patuxent River test team reported on the performance of the updated version of ALIS 1.0.3 in May 2013, and indicated that at least three of the four Category I deficiencies identified during initial testing remained open.
- Prior to the start of the Block 2B operational utility evaluation, the program must correct deficiencies in ALIS 1.0.3, finish development of ALIS 2.0, and integrate the propulsion module in ALIS 2.0.1, which is required for Marine Corps Initial Operational Capability (IOC). The Edwards test center plans to begin testing of ALIS 2.0 in April 2014 and ALIS 2.0.1 in September 2014. Delays in the release of ALIS 2.0 or 2.0.1 will add schedule risk to the Block 2B fleet release planned for mid-2015.
- The current Squadron Operating Unit (SOU) used by ALIS failed to meet the deployability requirement in the ORD due to the size, bulk, and weight of the current SOU design. To address the requirement, the program is developing a deployable version of the SOU, deemed SOU V2. It will support aircraft in the Block 2B, 3i, and 3F configuration, and is a critical delivery item for meeting Service IOC dates.

The Program Office has divided the SOU V2 development into multiple increments.
  - The first increment includes the capability to deploy and support the requirements for Marine Corps IOC. This increment will align hardware (SOU V2) and software (ALIS 2.0.1) releases to allow testing to begin at the Edwards flight test center in January 2015.
  - The second increment, currently unfunded, will address U.S. Air Force requirements for sub-squadron reporting capabilities and inter-squadron unit connectivity.
  - A third increment, also unfunded, plans to add decentralized maintenance capability, which will allow personnel to manage tasks with or without connectivity to the main SOU.
- To date, diagnostic system performance has failed to meet basic functional requirements, including fault detection, fault isolation, and false alarm rates. Due to the failure to meet these requirements, the program has discontinued the development of enhanced diagnostics (model-based reasoning) for the remainder of SDD. The program has initiated manual workarounds in the field, such as maintainer-initiated built-in tests and reliance on contractor support personnel, for more accurate diagnostics of system faults.

Joint Technical Data

- Development of Joint Technical Data (JTD) modules for the F-35A and F-35B is largely complete. Verification naturally lags behind development, but is progressing toward completion. Verification of modules requiring extensive intrusion into the aircraft is planned to be completed during depot-level modifications or opportunistic maintenance. The F-35C lags behind the other variants, but is proceeding quickly because of variant similarities. The chart below shows the status of JTD development and verification for each variant, propulsion, support equipment, and sustainable low observable (SLO) maintenance. Results exclude JTD for pilot flight equipment and JTD unique to LRIP aircraft (such as structural field repairs) that will not be needed for full-rate production aircraft. From October 2012 to October 2013, the Program Office verified 2,581 aircraft and 822 propulsion modules. Early in 2014, the primary focus in JTD verification will be weapons and stores.

	Data Modules Identified (as of Oct 2013)	Data Modules Completed	% Data Modules Completed	% Data Modules Verified
F-35A[1]	4,404	4,045	91.9%	81%
F-35B[1]	5,314	4,766	89.7%	76%
F-35C[1]	4,514	3,357	74.4%	55%
Propulsion	2,892	2,861	98.9%	94%
SE	2,241	489	21.8%	13%
SLO	1,362	291	21.3%	3%
Total	20,727	15,809	76.3%	64%

Note: 1. Includes field and depot-level JTD for Operations and Maintenance (O&M) for air vehicle only

- As stated earlier in the F-35 fleet availability section, aircraft maintenance personnel submit ARs to Lockheed Martin when the needed JTD is not available to troubleshoot or resolve a problem with an aircraft. The time maintenance personnel wait for resolution of these ARs contribute to aircraft non-availability (25-30 percent of the reported NMC time has been due to AR wait time).
  - Lockheed Martin prioritizes and responds to ARs through the Lightning Support Team, which is composed of Service and contractor personnel. The support has been fairly successful in responding to the most critical ARs with at least an interim solution in a timely manner, but because of manpower limitations, has been unable to handle the backlog of less severe ARs.
  - As of August 2013, 231 critical ARs remained open, while over 200 severe ARs were open. A critical AR addresses a deficiency which may cause major loss or damage to a system, or severe injury or possible death to personnel if not corrected. A severe AR addresses a deficiency which adversely affects operational safety, suitability, or effectiveness; however, a workaround is permitted.

F-35B Air-Ship Integration and Ship Suitability Testing

- The Navy deployed two F-35Bs to LHD-1 (USS *Wasp*) for two weeks in August 2013 to continue assessing shipboard suitability and integration. The Navy is continuing to analyze data from this deployment. Permanent modifications to the *Wasp* to prepare for JSF integration included:
  - Addition of transverse stiffeners to the underside of the flight deck for the two landing spots used by the F-35B and application of thermal non-skid material to the topside of the flight deck for one landing location. The Marine Corps applied the non-skid material to the other landing location before an earlier detachment to the *Wasp*.
  - Deck edge modifications, including the removal, replacement, relocation, and shielding of communications systems.
  - Added fire detection and alarming systems for the lithium-ion battery charging and storage area.
- Temporary alterations for the *Wasp* for this detachment include:
  - Lithium-ion battery charging and storage areas. The Marine Corps has not determined the final design of these areas.
  - Short take-off rotation line lights. Analysis of results will determine the precise location of these lights.
  - Addition of test equipment.
- The deployment met the primary objective of collecting data to support the development of a Block 2B operational flight envelope for take-offs and landings. The test team expanded the range of aircraft weight and center of gravity compared to that developed from the first deployment in 2011 and conducted operations in both day and night conditions. The test team completed 95 short take-offs and vertical landings, including forward and aft facing landings, and 17 night take-offs and landings during the deployment.

- The Marine Corps is developing solutions to a number of challenges in integrating the F-35B onto L-class ships:
  - Large-scale application of a thermal non-skid material to the flight deck in F-35B landing locations.
  - Modification of the flight deck structure to eliminate excess stress, which includes transverse panel breakers installed on the underside of the existing flight deck structure.
  - Design of separate charging and storage lockers for the lithium-ion batteries required for the JSF and new storage locker for pilot flight equipment, as the JSF helmet is larger and more fragile than legacy helmets.
  - New firefighting procedures in the event of a fire on the flight deck near aircraft carrying internal ordnance.
  - Understanding requirements for gun pod storage.
  - Conducting feasibility studies on the resupply of F-35B engines while underway, which could include a greater space allocation for engine storage aboard ship or through underway replenishment using a Navy system currently installed on one supply ship and scheduled for installation on CVN-78.
  - The Marine Corps has determined that new active noise reduction personal hearing protection is necessary for on-deck personnel because of the high level of engine noise. Noise damping materials and/or personal hearing protection may also be needed for below-deck personnel.

F-35C Air-Ship Integration and Ship Suitability Testing

- Although a number of air-ship integration issues are common to both CVN and L-class ships, such as lithium-ion battery storage, pilot flight equipment storage, need for new shipboard firefighting procedures, and high noise levels, some issues and their solutions are particular to aircraft carriers. The Navy has made progress in addressing some of these integration issues, but several challenges remain.
  - The program began testing its redesigned arresting hook system on a flight test aircraft in late CY13. The redesign was necessary after the original system failed to engage the cable and demonstrate sufficient load-carrying capacity. The arresting hook system remains an integration risk as the JSF development schedule leaves no time for new discoveries. Other risks include the potential for gouging of the flight deck after a missed cable engagement (due to an increase in weight of 139 pounds) and the potential for sparking from the tail hook across the flight deck because of the increased weight and sharper geometry of the redesigned hook.
  - The Navy is redesigning the cooling system in the Jet Blast Deflectors, which deflect engine exhaust during catapult launches, to handle JSF engine exhaust. The redesign will include improvements in side-cooling panels.
  - CVN-78 will receive the new Heavy underway replenishment (UNREP) system along with one resupply ship, but the Navy has delayed this system for eight years on other ships. This new UNREP system is the

only system capable of transporting the JSF engine and container while the carrier is underway.

- The JSF engine container was unable to sustain the required sudden drop of 18 inches (4.5 g's) without damage to the power module during shock testing. The Navy is redesigning the container to better protect this engine, but this is likely to result in an increase in container size and weight. The Navy estimates new container availability in late 2016.

- Engine noise is a potential risk to personnel on the flight deck and one level below the flight deck. The Navy has decided to procure active noise reduction personal hearing protection for on-deck personnel. Projected noise levels one level below the flight deck (03 level) will require at least single hearing protection. On most carriers this is a berthing area, but on CVN-78 this is a mission planning space; personnel wearing hearing protection in mission-planning areas will find it difficult to perform their duties. The Navy previously tested acoustic damping material in 2012 and is developing a model to optimize material placement.

- Storage of the JSF engine is limited to the hangar bay, which will affect hangar bay maintenance operations. The impact on the JSF logistics footprint is not yet known.

- Lightning protection of JSF aircraft while on the flight deck will require the Navy to modify nitrogen carts to increase capacity. Nitrogen is used to fill fuel tank cavities and inert aircraft at specified intervals while on deck.

Progress in Plans for Modification of LRIP Aircraft

- The Program Office and Services continued planning for modification of early LRIP aircraft to attain planned service life and the final SDD Block 3 capability.
  - Planning has focused on modifying aircraft in preparation for the Block 2B operational utility evaluation and Marine Corps IOC, both planned to occur in 2015.
  - Because operational test aircraft are to be production-representative, the Program Office must coordinate verification and approval of all modifications, the availability of docks at the aircraft depots as they open for operation, and the availability of long-lead aircraft parts needed for modifications with inputs from the Services on modification priority.
- The Program Office developed a modification and retrofit database that contains information for each entry on Service prioritization, when the modification will become part of the production line, which aircraft will require modification, whether unmodified aircraft are limited in performance envelope and service life or will require additional inspections, and operational test requirements and concerns.
- Modifications that do not require depot induction will be performed by depot field teams (who will travel to aircraft operating locations or to depots to work alongside depot teams) or by unit-level maintainers. The Program Office and Services adjudicate the location of all Block 2B modifications.

- Modifications to support the operational utility evaluation of Block 2B capability include:
  - Missions systems modifications, including those for Block 2B capability
  - Structural life limited parts, referred to as Group 1 modifications
  - STOVL Mode 4 operations modifications, which include a modification to the Three Bearing Swivel Module, which is required to allow STOVL aircraft to conduct unrestricted Mode 4 operations
  - Lightning certification, which includes OBIGGS modification (the lightning qualification of line-replaceable components and development of a system-level test still need to be completed before the aircraft modifications can proceed)
  - Support/training systems, which include the ALIS and pilot training device to support operational test aircraft
  - Other modifications, including those to vehicle systems, airframes, aircraft operating limitations, and weapons.
- The concurrency of production with development created the need for an extensive modification plan to ensure aircraft are available and production-representative for operational testing. The current modification schedule contains no margin and puts at risk the likelihood that operationally representative aircraft will be available for the Block 2B operational utility evaluation when it is currently planned by the Program Office to occur in 2015.

Recommendations

- Status of Previous Recommendations. The program and Services are satisfactorily addressing three of ten previous recommendations. The remaining recommendations concerning correction of the schedule in the TEMP, end-to-end ALIS testing, VSim validation, alignment of weapons test schedules with the Integrated Master Schedule, test of the redesigned OBIGGS system, reinstatement of the PAO shut-off valve, reinstatement of the dry-bay fire extinguisher system, and provision of a higher resolution estimate of time remaining for controlled flight after a ballistic damage event are outstanding.
- FY13 Recommendations. The program should:
  1. Ensure flight test timeline estimates for remaining SDD flight testing faithfully account for the historical growth in JSF testing, in particular for mission systems and weapons integration.
  2. Plan realistic rates of accomplishment for remaining weapons integration events; assure the events are adequately resourced from the planning phase through data analysis.
  3. Resource and plan SDD flight test to acquire the needed validation data for VSim.
  4. Track and publish metrics on overall software stability in flight test. The stability metrics should be "mission focused" and account for any instability event in core

or sensor processors, navigation, communication, radar, EOTS, DAS, or fusion display to the pilot.

5. Design and reinstate an effective fueldraulic shut-off system to protect the aircraft from fuel-induced fires. Recent testing has shown that this feature could protect the aircraft from threat-induced fire; this is also a critical flight safety feature.

6. Determine the vulnerability potential of putting 270-volt power on a 28-volt signal bus. Due to the unique electrical nature of the F-35 flight control system, the Program Office should thoroughly examine and understand this vulnerability before this aircraft becomes operational. The Program Office should successfully incorporate the wire harness design and the associated vulnerabilities in the F-35 vulnerability analysis tools.

7. Develop a plan to improve the Integrated Caution and Warning system to provide the pilot with necessary vulnerability information. The vehicle system should have the capability of detecting and reporting to the pilot any component ballistic damage (e.g., lift fan shaft) that could lead to catastrophic failure (e.g., upon attempt to convert to STOVL flight).

8. Track LO and non-LO repair times across the fleet and report them separately in monthly performance metrics. Separately track LO repairs due to inherent LO failures and due to facilitating other maintenance actions, and note the proportion of all LO repairs that are caused by facilitating other maintenance actions.

9. Plan to conduct the operational utility evaluation of Block 2B using comparative testing of the capabilities Block 2B provides relative to the capabilities provided by legacy aircraft. This approach was used to test the F-22, and is particularly critical for Block 2B operational testing because no detailed formal requirements for Block 2B performance exist.

www.ingramcontent.com/pod-product-compliance
Lightning Source LLC
Chambersburg PA
CBHW081825170526
45167CB00008B/3544